<u>Ra</u>

By Robert

An Easy Steps Math book

Copyright © 2014 Robert Watchman

All rights reserved.

No portion of this publication may be reproduced, transmitted or broadcast in whole or in part or in any way without the written permission of the author.

Other books in the Easy Steps Math series

Fractions
Decimals
Percentages
Ratios
Negative Numbers
Algebra
Master Collection 1 – Fractions, Decimals and Percentages
Master Collection 2 – Fractions, Decimals and Ratios
Master Collection 3 – Fractions, Percentages and Ratios
Master Collection 4 – Decimals, Percentages and Ratios

More to Follow

Contents

Introduction

This series of books has been written for the purpose of simplifying mathematical concepts that many students (and parents) find difficult. The explanations in many textbooks and on the Internet are often confusing and bogged down with terminology. This book has been written in a step-by-step 'verbal' style, meaning, the instructions are what would be said to students in class to explain the concepts in an easy to understand way.

Students are taught how to do their work in class, but when they get home, many do not necessarily recall how to answer the questions they learned about earlier that day. All they see are numbers in their books with no easy-to-follow explanation of what to do. This is a very common problem, especially when new concepts are being taught.

For over twenty years I have been writing math notes on the board for students to copy into a note book (separate from their work book), so when they go home they will still know how the questions are supposed to be answered. The excuse of not understanding or forgetting how to do the work is becoming a thing of the past. Many students have commented that when they read over these notes, either for completing homework or studying for a test or exam, they hear my voice going through the explanations again.

Once students start seeing success, they start to enjoy math rather than dread it. Students have found much success in using the notes from class to aid them in their study. In fact students from other classes have been seen using photocopies of the notes given in my classes. In one instance a parent found my math notes so easy to follow that he copied them to use in teaching his students in his school.

You will find this step-by-step method of learning easier to follow than traditional styles of explanation. With questions included throughout, you will gain practice along with a newfound understanding of how to complete your calculations. Answers are included at the end.

Chapter 1

Ratio Basics

A ratio is simply a comparison of two or more amounts. This can include whole numbers, fractions, decimals or percentages. A good working knowledge of these areas will be helpful in understanding ratios. (See the Easy Steps Math Master Collection 1 book)

For example at a conference there are 40 men and 100 women. The ratio of men to women is 40 to 100. This can also be written as 40:100, and can be simplified, just like fractions, to 2:5. A ratio can also be written as a fraction. The ratio of men to women can also be written as $\frac{40}{100}$ or $\frac{2}{5}$.

One important aspect of ratios is the order that the numbers are written. For instance, in the example above the ratio of 'men to women' is 2:5. If the ratio of 'women to men' is required then this is 5:2. **The wording must match the ratio figure otherwise errors will occur.**

Another important aspect is that the units of a ratio must be the same. An example of this is meters and kilometers. You must choose one (usually the smaller unit) and then state your ratio. So if your question is *state the ratio of 5 kilometers to 300 meters in shorthand form*, you must change kilometers to meters (5 km = 5000 m) first, and then state the ratio.

Therefore 5 km to 300 m
= 5000 m to 300 m
= 50 m to 3 m

Rewrite the following using the **shorthand ratio** symbol (:), simplify the ratio where necessary.

a) 13 books to 25 books

b) 10 babies to 23 babies

c) 16 apples to 19 apples

d) 4 kilometers to 9 kilometers

e) 20 cm to 41 cm

f) 29 minutes to 2 hours

g) 14 cm to 3 mm

h) 9 hours to 4 days

i) 4 km to 827 m

j) 18 kg to 2343 g

Chapter 2

Simplifying Ratios

As mentioned above, ratios can be simplified just like fractions. You can only multiply or divide both sides of the ratio by the same amount.

If either or both parts of the ratio are a fraction or a decimal, then they will need to be multiplied out to whole numbers.

Examples:

a) *Simplify* $14:49$

Step 1. Decide which number goes into both sides of the ratio without a remainder i.e. the largest common divisor. In this case it is 7.

Step 2. Divide both sides of the ratio by 7.

$$=14 \div 7 : 49 \div 7$$

$$=2:7$$

So the ratio $14:49$ is the same as $2:7$.

b) *Simplify* $3\frac{1}{2} : 2$

As the first part of the ratio is a mixed number, this will need to be changed to a whole number. Multiplying both sides by the same number will do this.

Step 1. Change the fraction in the ratio to improper fractions.

$\frac{7}{2} : \frac{2}{1}$

Step 2. Multiply both sides by the appropriate number. In this case it is 2. The reason for choosing 2 is that it is the product of the two denominators.

$= \frac{7}{2} \times \frac{2}{1} : \frac{2}{1} \times \frac{2}{1}$

$= \frac{14}{2} : \frac{4}{1}$

Step 3. Simplify

$= 7 : 4$

So the ratio $3\frac{1}{2} : 2$ is the same as $7 : 4$.

Occasionally there will be fractions in both parts of the ratio.

E.g.

c) *Simplify* $2\frac{1}{4} : 5\frac{2}{3}$

Step 1. Change the fractions in the ratio to improper fractions

$\frac{9}{4} : \frac{17}{3}$

Step 2. Multiply both sides of the ratio by the appropriate number. The product of both denominators is 12. So multiply by 12.

$=\frac{9}{4}\times\frac{12}{1} : \frac{17}{3}\times\frac{12}{1}$ cross simplify before you multiply.

$=\frac{9}{\cancel{4}_1}\times\frac{\cancel{12}^3}{1} : \frac{17}{\cancel{3}_1}\times\frac{\cancel{12}^4}{1}$

$=\frac{27}{1} : \frac{68}{1}$

$=27 : 68$

So the ratio $2\frac{1}{4} : 5\frac{2}{3}$ is the same as $27 : 68$

d) *Simplify* 2.5:4.2

As this ratio includes decimals, it will need to be converted to whole numbers before it is simplified. The easiest way to convert decimals to whole numbers is to multiply by a multiple of 10 (see the Easy Steps Math Decimals book).

Step 1. Multiply both sides of the ratio by a multiple of 10. In this case it is 10.

2.5 x 10 : 4.2 x 10

= 25:42

So the ratio 2.5:4.2 is the same as 25:42

Simplify each of the following.

a) 4:16

b) 51:85

c) 55:99

d) 32:88

e) 300:5000

f) 3 km : 15 km

g) 10 cars : 45 cars

h) 9 cm : 3 mm

i) 9 months : 1 year

j) $4.50 : 45 cents

k) $5 : 1\frac{1}{2}$

l) $7\frac{3}{4} : 5$

m) $12 : 3\frac{2}{7}$

n) $8\frac{1}{6} : 2$

o) $3\frac{1}{2} : 4\frac{3}{4}$

p) 2.3:4.5

q) 6.8:5.9

r) 7.2:9.4

s) 4.65:8.1

t) 5.21:3.45

Chapter 3

Proportion

When two or more **ratios are in proportion**, it means they are equivalent to each other.

For example,

4:5 is in proportion to 16:20 and 1:3 is in proportion to 8:24 and 2:7 is in proportion to 16:56.

These ratios can be written like so:

4:5 = 16:20, 1:3 = 8:24, and 2:7 = 16:56

or like so:

$$\frac{4}{5}=\frac{16}{20},\ \frac{1}{3}=\frac{8}{24},\ \text{and}\ \frac{2}{7}=\frac{16}{56}$$

We write ratios as fractions, so we can check if they are in proportion. We do this by using a method called **cross-multiplication.** This is where the numerator of one fraction is multiplied with the denominator of the other fraction and vice versa. The answers for both must be the same.

For instance, to check if these ratios are in proportion, we cross-multiply (multiply in the direction of the arrows as shown).

$$\frac{4}{5} \nwarrow\!\searrow \frac{16}{20} \ \text{and}\ \frac{4}{5} \swarrow\!\nearrow \frac{16}{20}$$

Multiplying 4 and 20 gives us 80, (4 x 20 = 80) and multiplying 5 and 16 also gives us 80, (5 x 16 = 80). Therefore the statement that the ratios $4:5 = 16:20$ are the same is true. They are in proportion.

Are the following ratios in proportion? Answer True or False for each one. (Use the cross-multiplication method).

a) $2:3$ and $8:12$

b) $4:7$ and $8:14$

c) $7:9$ and $21:25$

d) $3:8$ and $12:32$

e) $14:16$ and $5:9$

f) $11:12$ and $7:8$

g) $\frac{13}{15}$ and $\frac{6}{7}$

h) $\frac{8}{9}$ and $\frac{24}{27}$

i) $\frac{3}{5}$ and $\frac{6}{8}$

j) $\frac{21}{18}$ and $\frac{49}{42}$

Being able to work out proportions allows us to keep amounts in the correct ratios.

For example in order to mix chemicals in a lab you need to know what proportions you are mixing. Knowing this, you can mix the equivalent amounts on a larger or smaller scale. For instance, an industrial chemist has come up with a new improved formula for detergent. He knows that he needs to mix two specific chemicals, A and B in the ratio $7{:}9$. He goes out to the factory floor and sees that he has none of chemical A and 63 liters of chemical B. How much of chemical A will he need to buy to be able to make the new detergent? To work this out you need to do the following:

A:B = 7:9

We know he has 63 liters of chemical B

A:63 = 7:9

Written as a fraction you would have

$$\frac{A}{63}=\frac{7}{9}$$

Cross multiply

$$A\times 9=7\times 63$$

$A\times 9=441$ (divide both sides of the equals sign by 9)

$$A=49$$

So the chemist needs 49 liters of chemical A.

Example 2.

Find the value of x in the following proportions: $\frac{x}{5}=\frac{10}{25}$

Step 1. As the question is already written as a fraction, there is nothing to do here.

Step 2. Cross multiply

$$x \times 25 = 5 \times 10$$

$$x \times 25 = 50$$

Step 3. Divide both sides of the equals sign by the highest common divisor (this is 25).

$$\frac{x \times 25}{25} = \frac{50}{25}$$

$$\frac{x \times \cancel{25}}{\cancel{25}} = \frac{\cancel{50}^{2}}{\cancel{25}}$$

Step 4. Simplify to obtain the answer.

$$x = 2$$

Example 3.

Find the value of a.

$4:3 = 12: a$

Step 1. Rewrite the question in fraction form

$$\frac{4}{3} = \frac{12}{a}$$

Step 2. Cross multiply

$$4 \times a = 3 \times 12$$

$$4 \times a = 36$$

Step 3. Divide both sides of the equals sign by the highest common divisor (this is 4).

$$\frac{4 \times a}{4} = \frac{36}{4}$$

$$\frac{\cancel{4} \times a}{\cancel{4}} = \frac{\cancel{36}^{9}}{\cancel{4}}$$

Step 4. Simplify to obtain the answer.

$$a = 9$$

Another way to do this question is to turn **<u>both</u> fractions upside down** so the unknown (the pronumeral) is on the top.

<u>Do not try this with any other fraction work. This can only be used for ratios that are in proportion.</u>

So these ratios $\frac{4}{3} = \frac{12}{a}$ can be changed to $\frac{3}{4} = \frac{a}{12}$ before you cross multiply. The answer will end up being the same.

Note that the value of the unknown is not always a whole number. Sometimes it is necessary to give the answer as a decimal.

Try these questions using the same method. Find the value of the pronumeral for each.

a) $x:2=5:10$

b) $x:6=12:18$

c) $\frac{x}{9}=\frac{2}{3}$

d) $3:x=18:24$

e) $\frac{7}{x}=\frac{12}{48}$

f) $\frac{10}{x}=\frac{9}{45}$

g) $3:7=x:35$

h) $\frac{12}{10}=\frac{x}{5}$

i) $\frac{8}{12}=\frac{x}{9}$

j) $40:60=3:x$

Chapter 4

Comparing Ratios

At times you will need to **compare ratios**, that is you will need to determine whether ratios are larger or smaller than each other or if the are equal to each other.

In order to compare ratios, you will need to rewrite them as fractions, and then make the denominators the same so as to determine which is larger or smaller.

For example,

Which is the larger ratio 4:5 or 5:7?

Step 1. Rewrite the ratios as fractions

$$\frac{4}{5} \quad \frac{5}{7}$$

Step 2. Make the denominators the same by finding the lowest common denominator (see the Easy Steps Math Fractions book)

$$\frac{28}{35} \quad \frac{25}{35}$$

Since the denominators are the same, the larger fraction will be the one with the larger numerator. In this case $\frac{28}{35}$ is the larger fraction.

Step 3. Insert a > or < sign between the fractions.

$$\frac{28}{35} > \frac{25}{35}$$

Therefore 4:5 is the larger ratio.

Compare the following pairs of ratios and decide which is larger.

a) 1:5, 2:5

b) 5:9, 7:9

c) 6:4, 11:8

d) 6:7, 12:14

e) 7:8, 2:2

f) 4:5, 5:6

g) 5:9, 7:11

h) 9:8, 6:5

i) 2:3, 9:12

j) 4:9, 5:12

Chapter 5

Increasing and Decreasing in a Given Ratio

A number can be increased or decreased by multiplying it by a given ratio (which has been turned into a fraction).

For instance, if you multiply any number by 1, that number does not change. However if you multiply that same number by a fraction, then the number would go up or down depending on the fraction. If the fraction is a proper fraction, then the number would go down. If the fraction is an improper fraction, then the number would go up. Consider the following examples.

1) $15 \times \frac{3}{5} = 9$ The number 15 is multiplied by a proper fraction, $\frac{3}{5}$, and the result went down to 9.

2) $15 \times \frac{5}{3} = 25$ The number 15 is multiplied by an improper fraction, $\frac{5}{3}$, and the result went up to 25.

3) $15 \times \frac{3}{3} = 15$ The number 15 is multiplied by 1, $\frac{3}{3}$, and the result stayed the same.

Therefore since a proper fraction is a ratio less than one, it decreases the value of the number when multiplied, and an improper fraction is a ratio greater than one, it increases the value of the number when multiplied.

For more information on how to multiply fractions refer to the Easy Steps Math Fractions book.

Complete the following:

a) Increase $45 in the ratio 6:5

b) Increase 120 kg in the ratio 9:6

c) Increase 90 L in the ratio 4:3

d) Increase 20 cm in the ratio 100:1

e) Decrease $150 in the ratio 3:5

f) Decrease 110 m in the ratio 7:10

g) Decrease 55 L in the ratio 1:100

h) Decrease 49 kg in the ratio 3:7

i) Change $55 in the ratio 5:11 and state whether it has increased or decreased.

j) Change 230 L in the ratio 15:10 and state whether it has increased or decreased.

Chapter 6

Dividing in a Given Ratio

If two business partners put money into a business in the ratio 7:3, then it would be logical that all profits would be divided in the same ratio and not equally. Assuming the business made $400,000 profit in the year; you would follow these steps to determine how much each partner would get.

Step 1. Add the two parts of the ratio together

$7 + 3 = 10$

Step 2. Take the total of the two ratios and divide the amount by this total

$400{,}000 \div 10 = 40{,}000$

Step 3. Multiply each part of the ratio by this answer above.

$7 \times 40{,}000 = 280{,}000$
$3 \times 40{,}000 = 120{,}000$

Therefore $400,000 divided in the ratio 7:3 is $280,000 : $120,000 which are the amounts the partners would get.

If the ratio has three parts the same thing is done.

E.g. Divide 160 in the ratio 3:1:4

Step 1. Add the parts of the ratio

$3 + 1 + 4 = 8$

Step 2. Divide the amount by the total of the two ratios.

$160 \div 8 = 20$

Step 3. Multiply each part of the ratio by this answer above.

$3 \times 20 = 60$
$1 \times 20 = 20$
$4 \times 20 = 80$

Therefore 160 divided in the ratio 3:1:4 is 60:20:80

(You can check your answer by adding these parts together. The result should be the amount in the question. So $60 + 20 + 80 = 160$, therefore correct)

Try these questions. Divided the amount in the ratio stated in brackets.

a) 60 (2:4)

b) 72 (2:6)

c) 42 (5:2)

d) 60 (8:2)

e) 48 (7:5)

f) 96 (2:10)

g) 400 (28:12)

h) 55 (9:2)

i) 390 (2:8:3)

j) 112 (4:1:3)

Chapter 7

Unit Rates/Unit Ratios

Some ratios cannot be simplified easily as they may end up as long decimals. These ratios can be converted to unit rates or unit ratios (they are the same thing). A unit rate has a denominator of 1. So $12:1$, $2.65:1$ and $\frac{1.25}{1}$ are example of unit rates. This format allows a comparison of the two figures.

For example, the ratio $12:1$ indicates that the first item being compared is 12 times larger than the second. The ratio $2.65:1$ indicates that the first item being compared is 2.65 times larger than the second, and the fraction $\frac{1.25}{1}$ indicates that the first item being compared is 1.25 times larger than the second. This system can be used to compare heights, lengths, mass, etc.

Assume that the height of two buildings is in the ratio $473:375$. This ratio is hard to visualize, so it can be changed to a unit rate, which allows a better understanding of the height comparison.

To change $473:375$ to a unit rate you would do the following:

Step 1. Make sure the smaller of the two numbers in the ratio is on the right (so $375:473$ would be switched around).

Step 2. Divide the two parts of the ratio by the smaller number

$473 \div 375 = 1.26$ (to 2 decimal places) and $375 \div 375 = 1$

or

$$\frac{473}{375} = 1.26 \text{ and } \frac{375}{375} = 1$$

Step 3. Rewrite the ratio with the decimal on the left and the 1 on the right.

1.26:1

Now we know that the larger of the two buildings is 1.26 times taller than the smaller building.

Write each of the following as unit rates. Give answers to 2 decimal places.

a) 15:14

b) 31:47

c) 733:525

d) 2357:2001

e) 652:865

f) 9841:5647

g) 354:568

h) 3485:5686

i) 787:354

j) 876:9860

Chapter 8

Scale Drawings

A scale drawing is an accurate drawing that represents a larger object or space, but is drawn to fit in a smaller area like a map or a blueprint. All the measurements are accurate but they are drawn according to the scale. The scale is written as a ratio to ensure accurate measurement and translation to the original size. The two parts of the scale ratio are the number 1 and the **scale factor.** The scale factor is the larger of the two numbers when the ratio is written with a 1 as the smaller number, as in 1:100,000. It is used for enlarging or reducing figures.

For example, a map may have a scale ratio of 1:100,000, which means that 1 cm on the map is 100,000 cm in actual size. Therefore the scale factor is 100,000. In 1:10, the scale factor is 10, in 1:100, the scale factor is 100, in 1:1000, the scale factor is 1000, etc. If necessary, you would simplify a ratio until it is in the form of a scale ratio.

Simplify the following to scale ratios and underline the scale factor. Remember to make units the same if needed.

a) 3:600

b) 7:14000

c) 9:900,000

d) 8:5600

e) 3 in : 3 yds

f) 2 mm : 50 cm

g) 4 yds : 1 mile

h) 5 cm : 1 m

i) 10 cm : 1 km

j) 1 mm : 1 m

To work out how far or how long a certain distance is, it is important to be able to do a conversion of the lengths on the diagram to real lengths. You would use the following method to do this.

In order to convert diagram length to real length, you multiply the diagram length by the scale factor.

Real Length = Diagram Length x Scale Factor

In order to convert real length to diagram length, you divide the real length by the scale factor.

Diagram Length = Real Length ÷ Scale Factor

As mentioned earlier, in order for any scale to be accurate the units must be the same. You cannot mix meters and kilometers, yards and miles, etc.

Complete the questions below using the rules for Real Length and Diagram Length above.

Use a scale of 1:100, to convert the following diagram lengths to real lengths. Give your answers in meters

a) 2 cm

b) 140 mm

c) 15cm

d) 12.9 cm

e) 26.5 cm

Use a scale of 1:100, to convert the following real lengths to diagram lengths. Give your answers in centimeters

f) 5.4 m

g) 23 m

h) 1.7 m

i) 2.7 m

j) 5 m

Multiplication Tables

To make calculations really easy, learn your multiplications tables. Here is a set of multiplication tables from 1 x 1 to 12 x 12 to help you if you need it.

1 x 1 = 1	2 x 1 = 2	3 x 1 = 3	4 x 1 = 4
1 x 2 = 2	2 x 2 = 4	3 x 2 = 6	4 x 2 = 8
1 x 3 = 3	2 x 3 = 6	3 x 3 = 9	4 x 3 = 12
1 x 4 = 4	2 x 4 = 8	3 x 4 = 12	4 x 4 = 16
1 x 5 = 5	2 x 5 = 10	3 x 5 = 15	4 x 5 = 20
1 x 6 = 6	2 x 6 = 12	3 x 6 = 18	4 x 6 = 24
1 x 7 = 7	2 x 7 = 14	3 x 7 = 21	4 x 7 = 28
1 x 8 = 8	2 x 8 = 16	3 x 8 = 24	4 x 8 = 32
1 x 9 = 9	2 x 9 = 18	3 x 9 = 27	4 x 9 = 36
1 x 10 = 10	2 x 10 = 20	3 x 10 = 30	4 x 10 = 40
1 x 11 = 11	2 x 11 = 22	3 x 11 = 33	4 x 11 = 44
1 x 12 = 12	2 x 12 = 24	3 x 12 = 36	4 x 12 = 48

5 x 1 = 5	6 x 1 = 6	9 x 1 = 9	10 x 1 = 10
5 x 2 = 10	6 x 2 = 12	9 x 2 = 18	10 x 2 = 20
5 x 3 = 15	6 x 3 = 18	9 x 3 = 27	10 x 3 = 30
5 x 4 = 20	6 x 4 = 24	9 x 4 = 35	10 x 4 = 40
5 x 5 = 25	6 x 5 = 30	9 x 5 = 45	10 x 5 = 50
5 x 6 = 30	6 x 6 = 36	9 x 6 = 54	10 x 6 = 60
5 x 7 = 35	6 x 7 = 42	9 x 7 = 63	10 x 7 = 70
5 x 8 = 40	6 x 8 = 48	9 x 8 = 72	10 x 8 = 80
5 x 9 = 45	6 x 9 = 54	9 x 9 = 81	10 x 9 = 90
5 x 10 = 50	6 x 10 = 60	9 x 10 = 90	10 x 10 = 100
5 x 11 = 55	6 x 11 = 66	9 x 11 = 99	10 x 11 = 110
5 x 12 = 60	6 x 12 = 72	9 x 12 = 108	10 x 12 = 120

7 x 1 = 7	8 x 1 = 8	11 x 1 = 11	12 x 1 = 12
7 x 2 = 14	8 x 2 = 16	11 x 2 = 22	12 x 2 = 24
7 x 3 = 21	8 x 3 = 24	11 x 3 = 33	12 x 3 = 36
7 x 4 = 28	8 x 4 = 32	11 x 4 = 44	12 x 4 = 48
7 x 5 = 35	8 x 5 = 40	11 x 5 = 55	12 x 5 = 60
7 x 6 = 42	8 x 6 = 48	11 x 6 = 66	12 x 6 = 72
7 x 7 = 49	8 x 7 = 56	11 x 7 = 77	12 x 7 = 84
7 x 8 = 56	8 x 8 = 64	11 x 8 = 88	12 x 8 = 96
7 x 9 = 63	8 x 9 = 72	11 x 9 = 99	12 x 9 = 108
7 x 10 = 70	8 x 10 = 80	11 x 10 =110	12 x 10 = 120
7 x 11 = 77	8 x 11 = 88	11 x 11 = 121	12 x 11 = 132
7 x 12 = 84	8 x 12 = 96	11 x 12 = 132	12 x 12 = 144

Answers

Using Ratio Shorthand

a) 13:25 b) 10:23 c) 16:19 d) 4:9 e) 20:41 f) 29:120
g) 140:3 h) 9:96 i) 4000:827 j) 1800:2343

Simplifying Ratios

a) 1:4 b) 3:5 c) 5:9 d) 4:11 e) 3:50 f) 1:5 g) 2:9 h) 30:1
i) 3:4 j) 10:1 k) 10:3 l) 31:20 m) 84:23 n) 49:12
o) 14:19 p) 23:45 q) 68:59 r) 36:47 s) 31:54 t) 521:345

Cross Multiplication (T or F)

a) T b) T c) F d) T e) F f) F g) F h) T i) F j) T

Find the Value of the Pronumeral

a) $x=1$ b) $x=4$ c) $x=4$ d) $x=6$ e) $x=28$ f) $x=50$
g) $x=15$ h) $x=6$ i) $x=6$ j) $x=4.5$

Comparing Ratios

a) 2:5 b) 7:9 c) 6:4 d) in proportion e) 2:2 f) 5:6 g) 7:11
h) 6:5 i) 9:12 j) 4:9

Increasing and Decreasing Ratios

a) $54 b) 180 kg c) 120 L d) 20 m e) $90 f) 77 m g) 550 ml h) 21 kg i) $25 (dec) j) 345 L (inc)

Dividing in a Given Ratio

a) 20:40 b) 18:54 c) 30:12 d) 48:12 e) 28:20 f) 16:80
g) 280:120 h) 45:10 i) 60:240:90 j) 56:14:42

Unit Rates

a) 1.07:1 b) 1.52:1 c) 1.40:1 d) 1.18:1 e) 1.33:1 f) 1.74:1
g) 1.62:1 h) 1.60:1 i) 2.22:1 j) 11.26:1

Scale Factor

a) 1:200 b) 1:2000 c) 1:100,000 d) 1:700 e) 1:36
f) 1:250 g) 1:440 h) 1:20 i) 1:10,000 j) 1:1000

Real Length/Diagram Length

a) 2 m b) 14 m c) 15 m d) 12.9 m e) 26.5 m f) 5.4 cm
g) 23 cm h) 1.7 cm i) 2.7 cm j) 5 cm

Conversion of Basic Units

Distance

1 km = 1000 m
1 m = 100 cm
1 cm = 10 mm

1 mile = 1760 yards
1 yard = 3 feet
1 foot = 12 inches

Area

1 km^2 = 1,000,000 m^2 = 100 hectares (ha)
1 ha = 10,000 m^2
1 m^2 = 10,000 cm^2
1 cm^2 = 100 mm^2

1 acre = 4840 yd^2
1 yd^2 = 9 $feet^2$
1 $foot^2$ = 144 $inches^2$

Mass

1 tonne (metric ton)= 1000 kg
1 kg = 1000 g

1 ton = 2000 lb (US) or 2240 lb (UK)
1 lb = 16 oz

Volume

1 Mega Litre (ML) = 1000 L
1 L = 1000 millilitres (ml)

1 gal = 4 quarts
1 quart = 2 pints
1 pint = 2 cups
1 cup = 8 fl oz

Glossary of Useful Terms

Cross multiplication is where the numerator of one fraction is multiplied with the denominator of the corresponding fraction and vice versa.

Sum refers to addition. The sum of two numbers is the answer of one number **plus** another number. E.g. the sum of 2 and 6 is 8, ($2 + 6 = 8$).

Difference refers to subtraction. The difference between two numbers is the answer of one number **minus** another number. E.g. the difference between 6 and 2 is 4, ($6 - 2 = 4$).

Product refers to multiplication. The product of two numbers is the answer of one number **times** another number. E.g. the product of 2 and 6 is 12, ($2 \times 6 = 12$).

Quotient refers to division. The quotient is the answer of one number being **divided** by another number. E.g. the quotient of 6 and 2 is 3 ($6 \div 2 = 3$).

Thank you for reading!

Dear Reader,

I hope you found this **Easy Steps Math – Ratios** book useful, either for yourself or for your children.

The **Easy Steps Math** series began as a set of math notes that I used in class for my students to copy from the board. I found that doing this helped the students in at least two ways. Firstly, with their homework, because they didn't forget how to do the work that was explained in class, and secondly, with their results, because they used the notes to study for tests and exams.

Where students in other classes were getting detentions for non-completion of homework, my students were getting homework done, their results were improving and they were enjoying math.

Students from other classes, even older students, were thanking me for my notes, as they were copying them from their peers because they found them so easy to follow and learn from. During a parent/teacher conference, one parent also thanked me because of how his child was able to easily learn the work, and that he, as a teacher, was using my notes in his classes, in his school.

This is when I realised that these notes would benefit many more students if they were published. Thus we are at this point.

I welcome any comments you have about this **Ratios** book. Tell me what you liked, loved, or even hated about it. I'd be happy to hear from you. You can email me at robwatchman@gmail.com

Finally, I would like to ask a favour. I would appreciate it if you would write a review of this book so that others can get an idea of how helpful it may be for them or their children. You would be aware that reviews are hard to come by because many readers don't go back to where they purchased their books.

So if you have the time, here is the link to my author's page on Amazon. You can find all of my other books here also: http://amzn.to/1rlW6gr

Thank you so much for reading the **Easy Steps Math – Ratios** book and for spending your time with me.

In Gratitude,

Robert Watchman

Made in the USA
Middletown, DE
27 May 2020

96076874R00024